Farming

DEBORAH ELLIOTT

Wayland

Titles in the Into Europe series
include:

Energy
Environment
Farming
Transport

Picture Acknowledgements:
Chapel Studios 24 (top); Eye Ubiquitous 13 (Jim Jenkins), 20 (left, Eva Miessler),
21 (Mike Southern), 22 (top, Julia Waterlow, bottom, Paul Seheult); Tony Stone
23, 27 (Julian Calder), 44 (Gary Irving); Topham Picture Library 11, 17, 26 (top),
29, 31 (both), 33 (right), 42; Wayland Picture Library 4 (left), 16, 19, 20 (right), 30
(both), 32, 34, 35, 37, 38, 41; Zefa contents page, 4 (right), 7, 9, 15, 24 (bottom), 26
(bottom), 28, 33 (left), 36, 39, 40, 43. All artwork is by Malcolm Walker.

Designed by Malcolm Walker

Text based on *Farming in Europe* in the Europe series published in 1991.

First published in 1993 by Wayland (Publishers) Limited
61 Western Road, Hove, East Sussex BN3 1JD

© Copyright Wayland (Publishers) Limited

British Library Cataloguing in Publication Data
Elliott, Deb
    Farming. - (Into Europe Series)
    I. Title II. Series
    630

ISBN 0 7502 0757 4

Typeset by Kudos
Printed and bound by Canale & C.S.p.A. in Turin, Italy

Contents

Farming today 4

The Alps 10

Norway, Sweden and Finland 14

The Netherlands 16

The Mediterranean 20

Eastern Europe 25

The European Community 30

The environment 38

The wetlands 41

Farming tomorrow 43

Glossary 46

More information 47

Index 48

Farming today

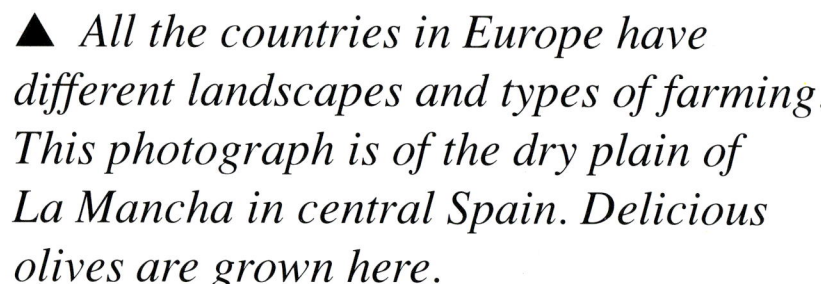

▲ *All the countries in Europe have different landscapes and types of farming. This photograph is of the dry plain of La Mancha in central Spain. Delicious olives are grown here.*

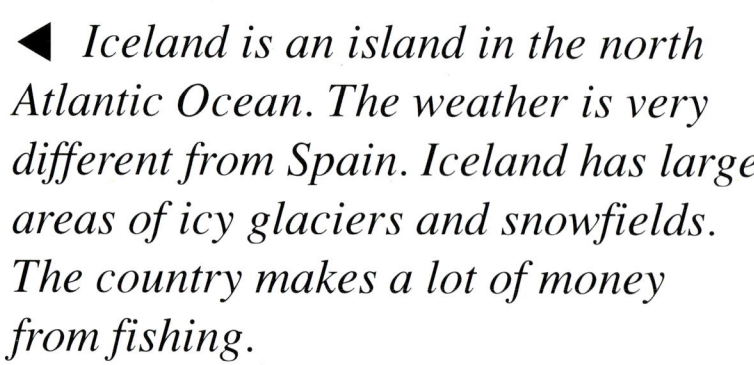

◄ *Iceland is an island in the north Atlantic Ocean. The weather is very different from Spain. Iceland has large areas of icy glaciers and snowfields. The country makes a lot of money from fishing.*

Farming is one of the most important industries in Europe. The crops and animals that farmers produce are sold to other countries. They also provide the food we need to live.

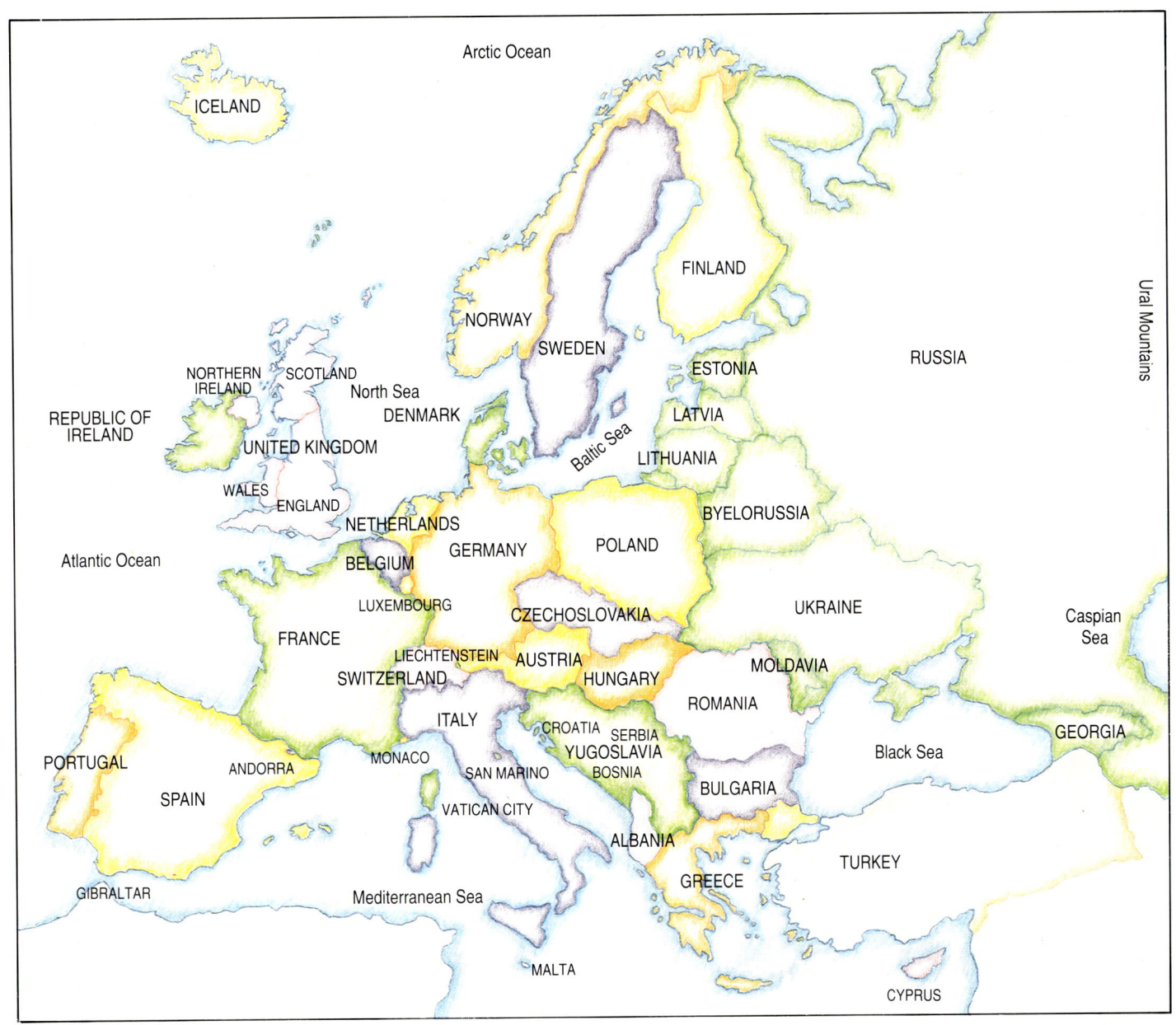

▲ *The countries in Europe. Can you find where you live?*

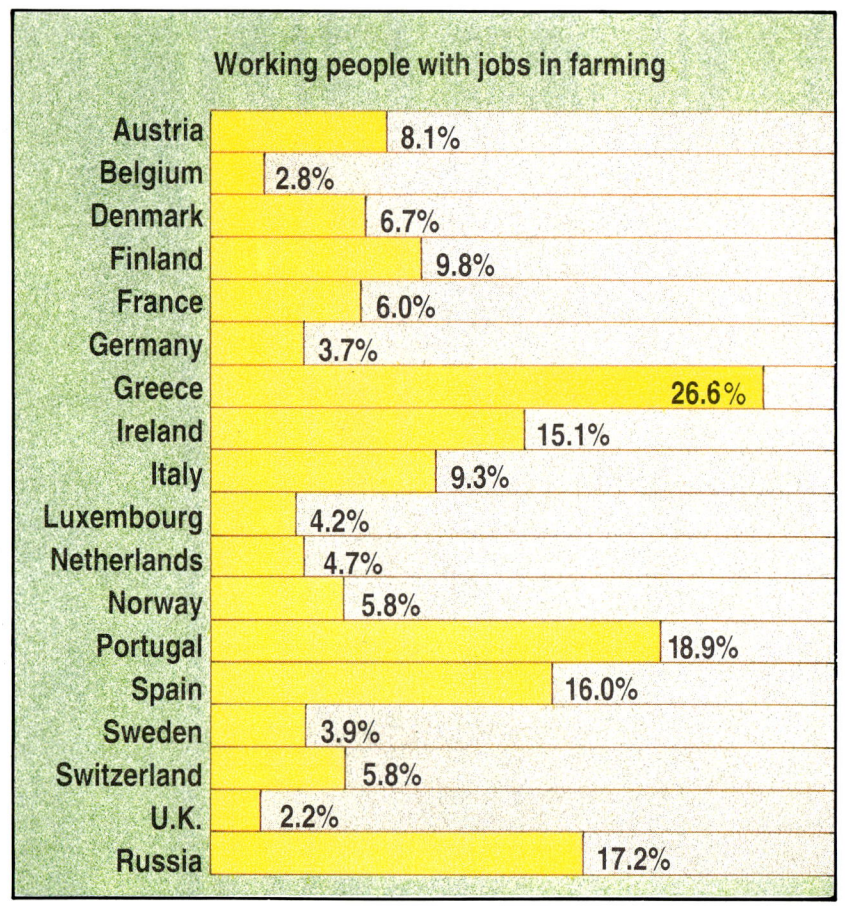

Working people with jobs in farming

Country	Percentage
Austria	8.1%
Belgium	2.8%
Denmark	6.7%
Finland	9.8%
France	6.0%
Germany	3.7%
Greece	26.6%
Ireland	15.1%
Italy	9.3%
Luxembourg	4.2%
Netherlands	4.7%
Norway	5.8%
Portugal	18.9%
Spain	16.0%
Sweden	3.9%
Switzerland	5.8%
U.K.	2.2%
Russia	17.2%

◀ *This chart shows the percentage of working people in Europe who work in farming.*

Can you see that farming is more important in some countries than others?

Farming has changed the environment in Europe. Forests have been cut down and water has been drained from wet, marshy land to make more fields for crops (see pages 38 - 42).

Farming provides the raw materials that are needed by lots of industries. For example, sugar cane or sugar beet is used in sugar refining. Cotton bolls are made into cotton for our shirts and shorts. Many people in Europe work in these industries.

This is a satellite photograph. It was taken in space looking down on Europe. It tells us what the weather will be like. ▶

▼ This chart shows the percentage of some foods produced in European countries. If a country produces more than 100 per cent, it can sell the extra. If it produces less, it has to buy food from other countries.

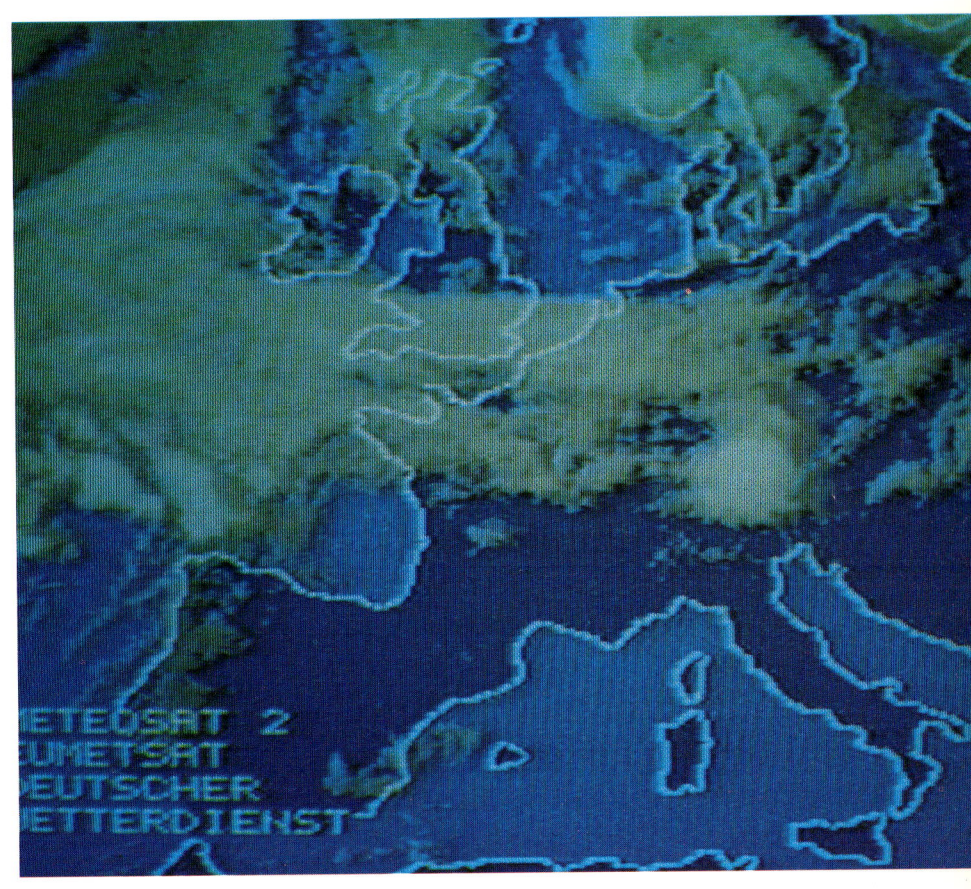

	Wheat	Potatoes	Sugar beet	Butter	Poultry	Vegetables
Austria	71	101	57	124	101	41
Finland	67	89	67	89	95	42
Portugal	104	97	112	47	86	141
Spain	115	89	131	79	98	165
Switzerland	96	104	110	165	97	94

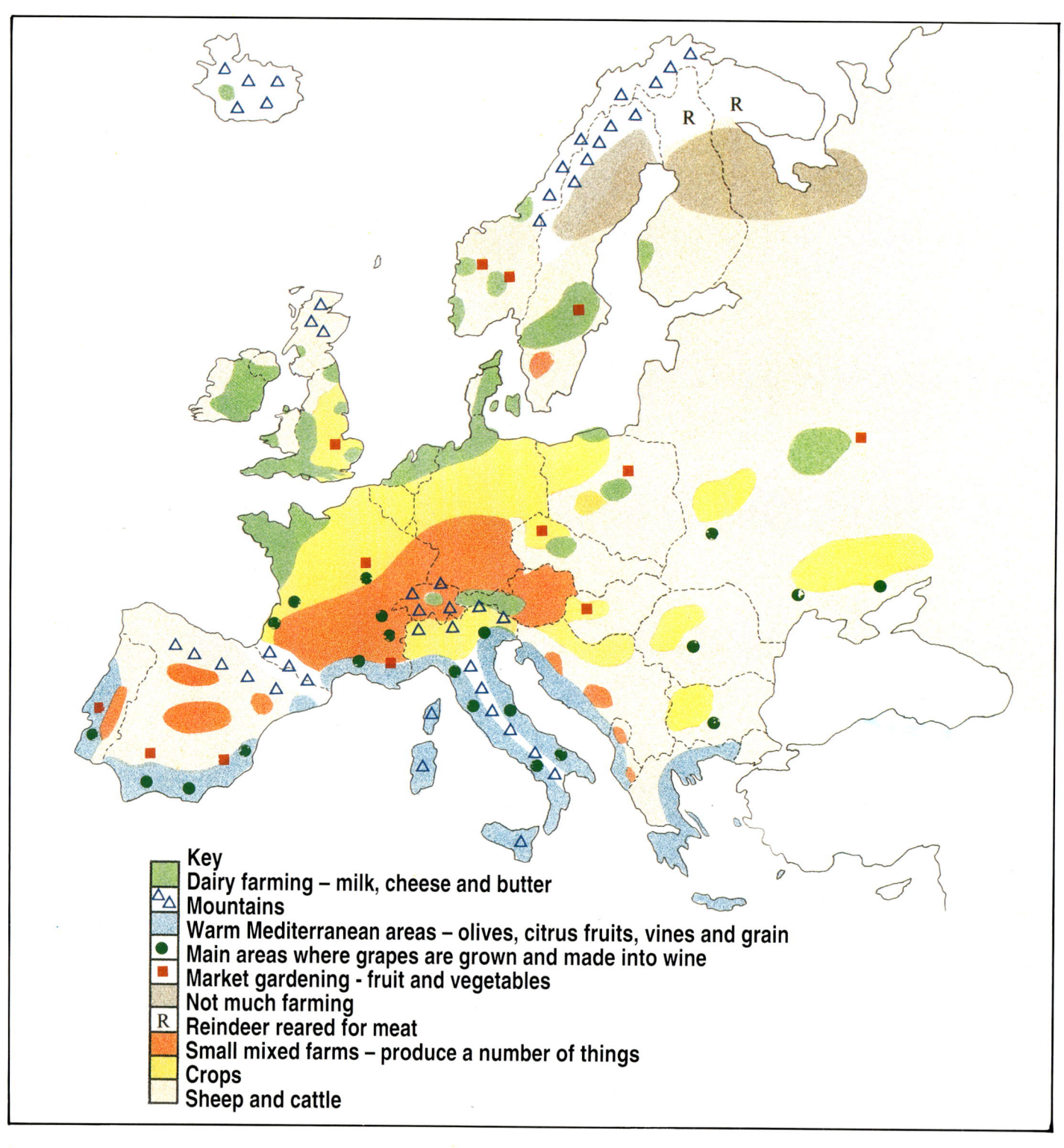

Key

- Dairy farming – milk, cheese and butter
- △ Mountains
- Warm Mediterranean areas – olives, citrus fruits, vines and grain
- ● Main areas where grapes are grown and made into wine
- ■ Market gardening - fruit and vegetables
- Not much farming
- R Reindeer reared for meat
- Small mixed farms – produce a number of things
- Crops
- Sheep and cattle

▲ *The different types of farming found in Europe.*

As you can see from the map opposite, all sorts of crops and animals are raised all over Europe. The type of farming depends on the height and shape of the land. For example, it is best to farm in the low, flat areas where the soil is usually rich and deep. Farming is much more difficult on high, steep land.

The climate also affects the type of farming. In wet, cloudy areas cattle, sheep and goats are reared. Crops are usually grown in drier, sunnier areas. Olives and citrus fruit (oranges, lemons and limes) can be grown in countries such as Portugal, southern Spain, Italy and Greece because the weather is so hot.

▲ *Most of the food found in shops comes from farms.*

The Alps

The Alps are a range of mountains. They stretch across France, Italy, Switzerland, Austria and Yugoslavia.

The Alps are well-known for their beauty. The icy glaciers and snow-covered peaks are perfect for skiing, walking and climbing. However, the steep valleys and rugged mountains have always made farming difficult.

In the countryside, farmers rear cattle for meat and for milk, cheese and butter. In the summer, the cattle are taken up the valley sides. Here they can graze on the grass. In the cold winter, the cattle are brought inside and fed hay.

Dairy farming in the Alps is not as popular as it used to be. Nowadays it is easier and cheaper for people to buy butter and cheese from other countries.

The photograph below shows cows being milked using modern machinery. Tubes are attached to the cows' udders and the milk is pumped out by machines. It is much quicker and easier than milking by hand.

▼ *Using machines to milk cows in the French Alps.*

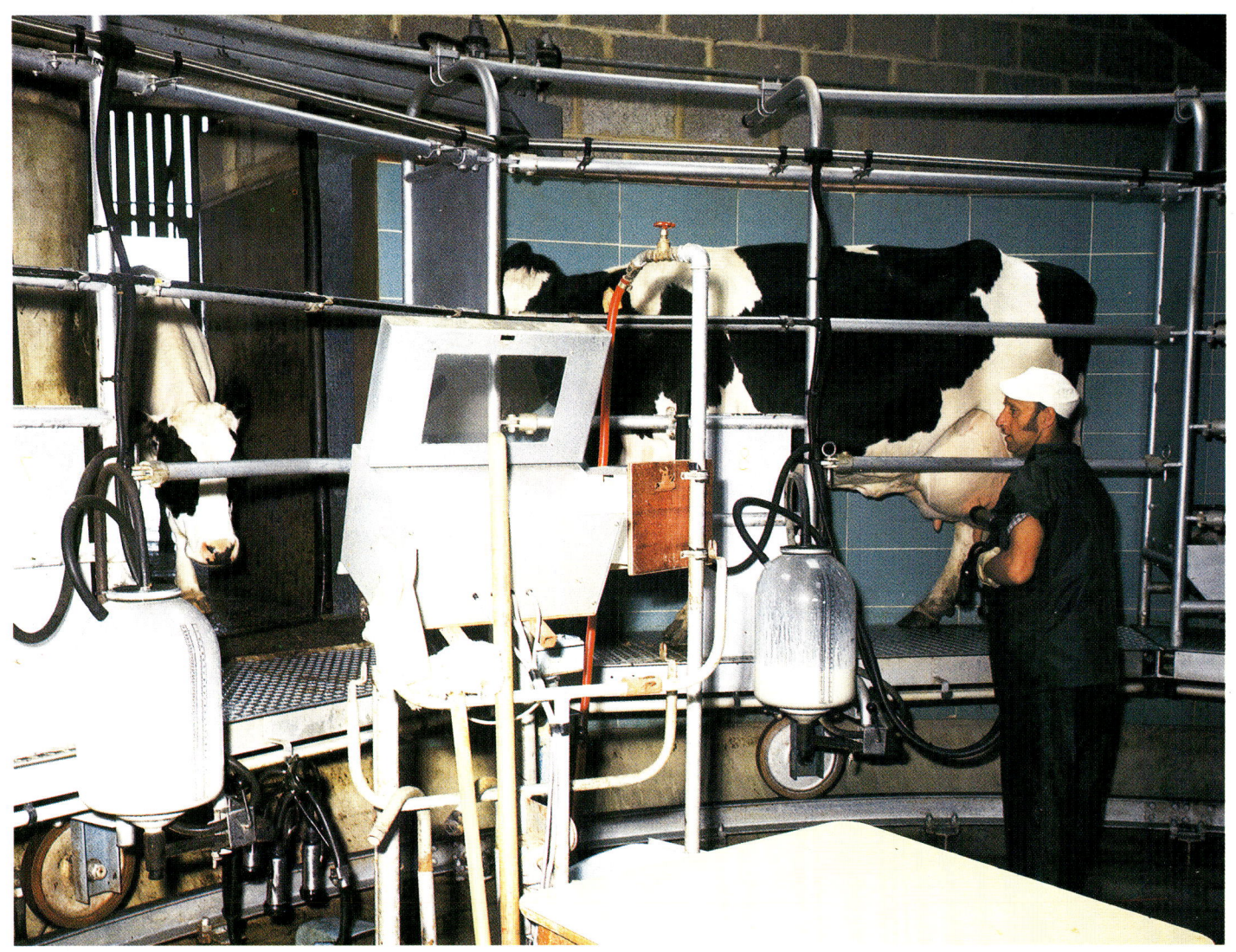

Switzerland

Much of Switzerland is made up of mountains, rock, forest and ice. In fact, only about one-quarter of the land is suitable for farming.

Most Swiss farmers rear cattle for milk. The milk is made into butter, fed to pigs and calves, and drunk by the Swiss. Some milk is made into cheese, such as Emmenthaler. It is also used to make the famous and delicious Swiss chocolate.

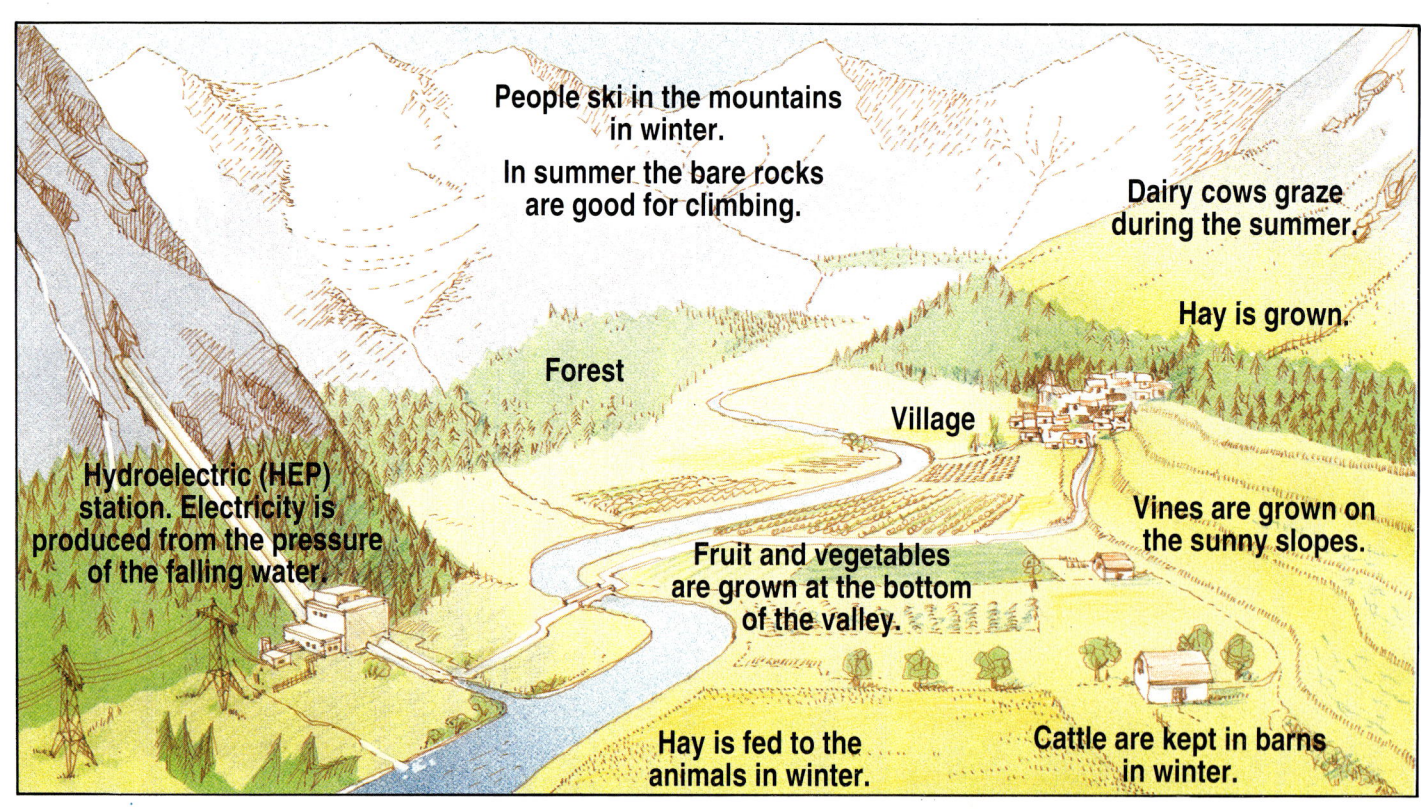

People ski in the mountains in winter.

In summer the bare rocks are good for climbing.

Dairy cows graze during the summer.

Hay is grown.

Forest

Village

Hydroelectric (HEP) station. Electricity is produced from the pressure of the falling water.

Vines are grown on the sunny slopes.

Fruit and vegetables are grown at the bottom of the valley.

Hay is fed to the animals in winter.

Cattle are kept in barns in winter.

▲ *Using a machine to gather up hay. Hay is fed to cattle in winter in the Alps.*

◄ *This drawing shows all the things you would find in a typical valley in the Alps.*

Norway, Sweden and Finland

Norway, Sweden and Finland lie to the very north of Europe. The climate of all three countries is very cold and the environment is very harsh. This makes farming difficult.

Norway is famous for its fjords and steep, snow-covered mountains. Fjords are narrow strips of sea-water that flow between high cliffs.

Dairy farming is popular in Norway. Most of the land is used to grow grass to feed the animals. Potatoes, fruit and wheat are grown in the south.

In the north of Sweden and Finland, winters are very long and cold. They last for over seven months of the year. The freezing climate makes farming almost impossible.

The climate is warmer in the south. Grass is the main crop. Barley, potatoes and sugar beet are also grown.

▲ *A view of a Norwegian fjord. Look at the steep mountains.*

The Netherlands

The Netherlands is a small, flat country on the coast of northern Europe. It is famous for its cheeses, canals and dykes.

Dykes are huge walls that are built to hold back the sea. Over one-quarter of the Netherlands is below sea-level. Dykes stop the sea from flooding the land.

▼ *A tulip field in the Netherlands. Flowers are a major part of Dutch farming.*

▲ *This land was once covered by the sea. Now a dyke keeps the sea back. Water gathers in ditches and is pumped into canals and rivers. Crops can be grown on the reclaimed land which is called a polder.*

Over 14 million people live in the Netherlands. Feeding so many people in such a small country is a big problem. Farmland has to be created. Land is taken (reclaimed) from the sea and from sandy heathland areas in the south and east of the country.

The Zuider Zee

The drawing below is of an area of reclaimed land in the Netherlands. It is called the Zuider Zee.

Before 1927 this whole area was part of the North Sea. A huge dyke was built to keep back the sea.

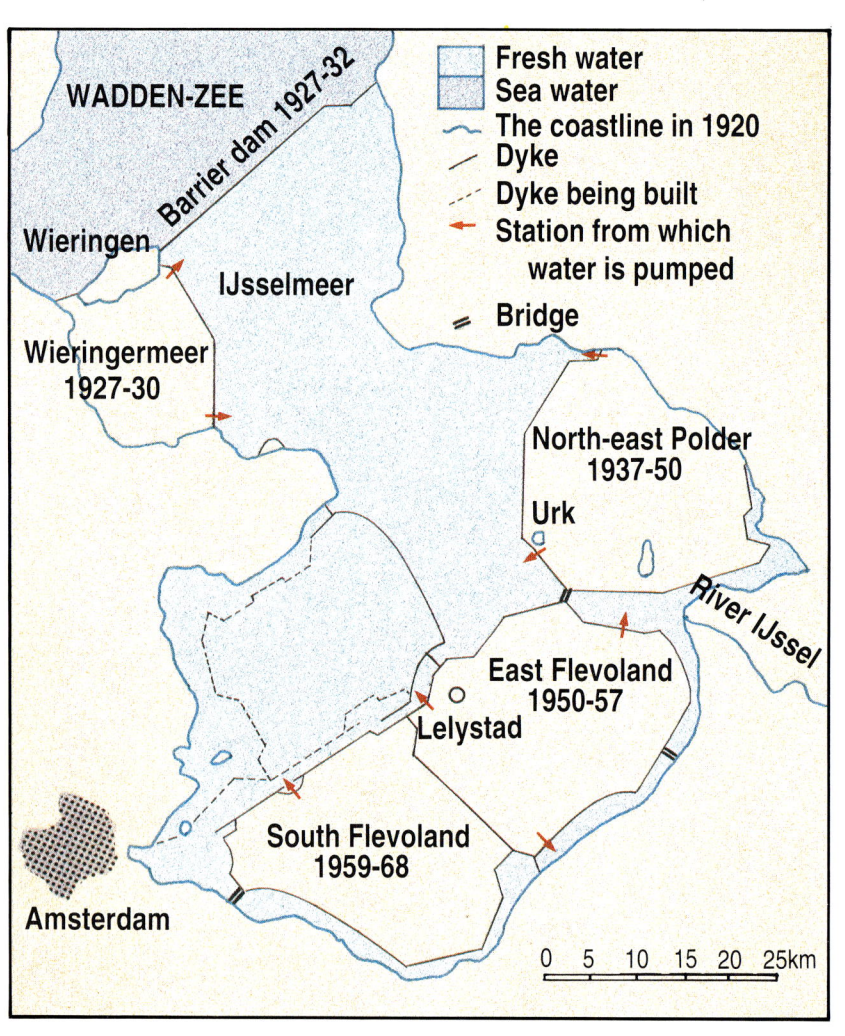

Fresh water
Sea water
~ The coastline in 1920
Dyke
Dyke being built
→ Station from which water is pumped
= Bridge

WADDEN-ZEE

Barrier dam 1927-32

Wieringen

IJsselmeer

Wieringermeer 1927-30

North-east Polder 1937-50

Urk

River IJssel

East Flevoland 1950-57

Lelystad

South Flevoland 1959-68

Amsterdam

0 5 10 15 20 25km

The new land is used mostly for farming. The area also has woodlands, lakes, beaches and nature reserves for animals and plants.

There are plans to take back more land from the sea. This will also be used for farming.

◄ *This is the Zuider Zee, a huge area of reclaimed land. Look how the area has changed since the 1920s.*

▲ *This building is the home of the Dutch government. It is in The Hague. All the decisions about farming in the Netherlands are made here.*

Meat and dairy farming are very important in the Netherlands. The cattle graze on grassland which covers almost two-thirds of the country. Wheat, sugar beet and potatoes are grown in the polders.

Bulbs are also an important part of farming in the Netherlands. They are sold in the rest of Europe and in the USA and Canada. Bulbs are the short, thick stems of some plants. They have leaves that store food for the plants. Bulbs are planted under the ground and flowers grow from them.

The Mediterranean

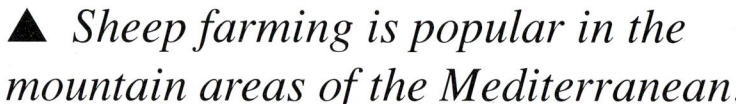

▲ *Sheep farming is popular in the mountain areas of the Mediterranean.*

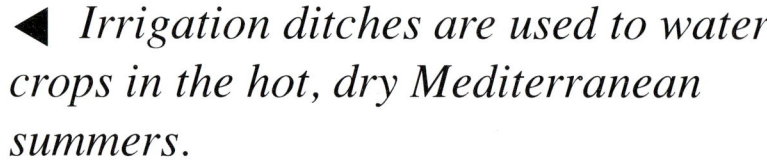

◄ *Irrigation ditches are used to water crops in the hot, dry Mediterranean summers.*

20

The Mediterranean is the name given to the countries in southern Europe - Spain, Italy, Greece and Portugal.

The dry summers in the Mediterranean make farming difficult. However, farming is one of the most important industries in the area.

Farmers grow figs, olives, peaches, pears and vines. Maize, rice, potatoes, tobacco, tomatoes, olive oil and wine are also produced and sold all over the world.

▲ *Crops are grown on this hillside terrace in Monterosso in Italy.*

◀ *Cotton is grown in this river valley in Greece. It is also grown in parts of Spain and Italy.*

Irrigation ditches provide water during the dry summer months.

Farmers in the Mediterranean use all sorts of different farming methods. Some use simple and traditional methods. The richer farmers use modern, high-tech machines.

You've probably heard of Chianti wine. It is famous all over the world.

The grapes that are used to make the wine are grown in vineyards in the Chianti region of Tuscany in Italy. ▶

Machines like this, which is being used to harvest grain, are becoming popular with farmers in the south of Italy. ▶

Farmers in the north of Italy use fertilizers and chemicals to grow huge amounts of fruit and vegetables. The summers in the north are hot, but there are also lots of rainstorms. These provide enough water to irrigate a large area of land.

In the south, however, summers are so hot and dry that farming is difficult. Many farmers are poor and have to rely on old-fashioned methods of farming.

◀ *Dairy farming is not popular in all Mediterranean countries. There are not enough crops that can be used to feed the cattle.*

In northern Italy, however, dairy farming is important. Cattle are kept inside and fed on hay and crops that are grown on irrigated land.

◀ *Farmers in the south of Italy have been given money by the government. They are planting new crops and trying different farming methods.*

This field of sunflowers in Sorrento, in southern Italy, will make a lot of money.

Eastern Europe

Russia is one of the countries in the east of Europe. Other countries like the Ukraine and Poland are close to Russia. You can find these countries and others mentioned in this chapter on the map on page 5.

Until lately, the governments in these countries told farmers what crops to grow and what animals to keep.

Russia is famous for its very big farms, called Collective farms. All the land is owned by the government. Workers are paid a share of the money made from sales of the crops or animals.

The government also own some farms, called State farms. Workers are paid a wage just as if they worked in a factory. The farms have their own schools, supermarkets and cinemas.

◄ Sugar beet is an important crop in Russia. Women usually help with the harvest.

Many Russian farms have factories where flour, sugar and milk can be produced.

Russian farms do not have modern equipment. This means that the harvest is often poor. In bad years, people who live in cities like Moscow have very little food.

There is not enough meat in Russia, so farmers are trying to rear more animals like these pigs. ►

There are a few large farms in the south of Russia and the Ukraine. Here huge combine harvesters are working across the enormous fields of wheat. ▶

Most Russians own a very small plot of land close to their homes. They grow vegetables, or keep a few chickens or a pig on the land.

Only 10 per cent of the land in Russia is suitable for farming. The rest is too cold, too dry or too mountainous.

Hungary, Czechoslovakia, Romania, Bulgaria and Albania all have Collective and State farms. In the past these farms have not produced enough food for everyone. Now they are being broken up into smaller farms which people can buy.

In Poland, Serbia, Croatia and Bosnia people have always owned their farms. However, the farms do not have enough modern tractors and other machines.

▼ *Horses are still used to pull hay wagons in much of eastern Europe. Lots of other farming jobs are also done by animals or people and not by machines.*

▲ *Rich fields of grass and clover are used to feed these beef cattle in Hungary.*

Farming in many of the countries in eastern Europe has changed a lot over the past few years. New farmhouses, new barns and even factories have been built.

However, despite all the changes there is still not enough food in the towns. Long queues of people outside food shops is a common sight.

Farmers are being encouraged to buy their own land. It is hoped that this will help them to produce more food. But farmers still need modern machines.

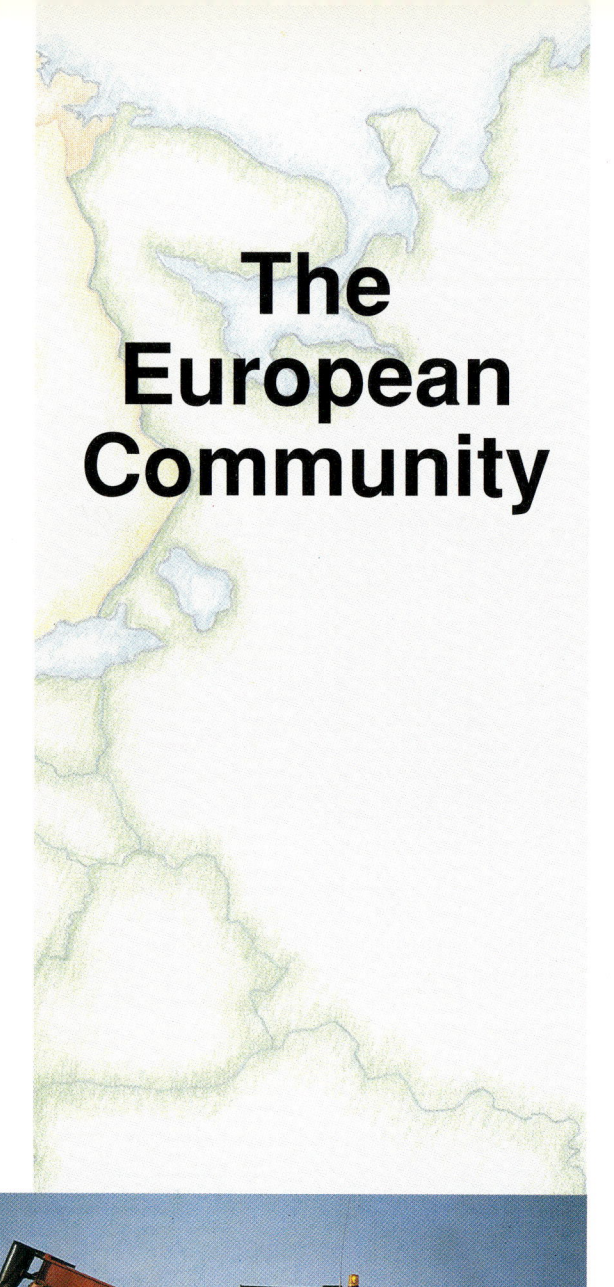

The European Community

The European Community is also known as the EC. It is an organization of twelve European countries who agree on some common policies. You can see which countries belong to the EC on the map on page 45.

The photograph on the right is of the EC headquarters in Brussels in Belgium. Here, in 1962, an agreement called the Common Agricultural Policy, or CAP, was made between the countries in the EC.

The CAP helps farmers in EC countries. For example, all the governments set prices (usually high) for the farmers' produce. It means that most farmers always make a profit. The CAP encourages farmers to use modern methods and to try new crops. A fund has been set up which gives farmers money for machinery and fertilizers.

◄ *In 1990, French farmers protested about changes that had been made to CAP policies. They let loose hundreds of sheep around the Eiffel Tower in Paris.*

The CAP has been good for farming in the EC. However, it has also caused problems. The fixed prices have meant that food in shops is expensive. The environment has been damaged by some of the chemicals used by farmers. Also, as farms have got bigger, hedgerows and wildlife have had to make way for larger fields.

Because farmers are given a fixed high price for their crops, many produce too much. This has led to food 'lakes' and 'mountains'.

EC countries produce too much butter, beef, milk and tomatoes. Some is stored in case of a bad harvest in the future. But most of the food is simply destroyed. ►

Ireland

This photograph is of a cattle farmer in the west of Ireland.

Ireland is very rural. Much of the land is made up of countryside and there is not much industry. The mild climate and fertile soils are good for farming.

Ireland joined the EC in 1973. Irish farmers have been helped by the CAP. They were encouraged to try more dairy farming. Now Irish farmers produce 5 billion litres of milk a year.

Irish farmers raise lambs for meat and grow cereals. These make a lot of money thanks to the high prices set by the EC.

Farmers in France use different methods. Many use modern machines. Others, like the farmer in the photograph, keep to traditional methods.

France produces most of its own food. It is famous for wheat, barley, maize, vines and olives. French farmers also raise sheep, cattle and poultry (chickens, ducks and turkeys) for meat.

Rows and rows of lavender are growing in this field in France. It must smell wonderful.

French farmers also grow delicious fruits and vegetables, such as apples, peaches, tomatoes and lettuce.

Britain

Only a small number of British people work in farming.

Farmers in Britain use modern machinery, such as combine harvesters.

The British government has asked farmers to produce more food. So, many areas of grassland have been ploughed and planted with seeds. Also, farmers use fertilizers and pesticides on their crops.

Many people are worried about food that has been sprayed with chemicals. They believe it makes some people ill.

This man is using a machine to **churn** *cream into butter in Denmark.* ▶

The countryside in Denmark is flat and the soils are not fertile. Danish governments have spent lots of money trying to make conditions better for farming.

In the past, farmers in Denmark grew mainly wheat and barley. However, wheat could be bought more cheaply from the USA and Canada. So many farmers changed to dairy and pig farming.

Milk is fed to the pigs and the cream is used to make cheese and butter.

◄ Pig farming is very important in Denmark. The pigs are reared for bacon which is sold all over the world. Have you ever tasted Danish bacon?

Danish farmers belong to co-operatives. These are groups of people who join together to help each other. They can buy large amounts of equipment, fertilizers and animal feed at cheap prices.

The co-operatives also collect and sell the milk and bacon. This means that customers can be sure that Danish products are always good.

Some of the co-operatives have set up their own banks. These banks lend money to farmers to buy machinery. They also give farmers free advice on new farming methods.

Farming is quite different in the east and west of Germany. Farms in the west are much smaller, more modern and make more money.

Most farmers in the west of the country only work part-time. They usually have other jobs in nearby towns.

Crops are grown on over half the farmland in the west. These include wheat, barley, maize, sugar beet and potatoes. Most of the crops are fed to cattle and pigs which are reared for their meat. Pork and German sausage, called *wurst*, are very popular.

◄ *This is a vineyard in the Rhineland – an area in the west of Germany.*

Grapes from the sunny slopes of the Rhine and Moselle valleys are turned into wine.

▲ *Before 1990, most farms in the east of Germany were owned by the government and everyone in the country. In October of that year, East and West Germany joined together to become one country. Now farmers have their own plots of land.*

The CAP (see page 30) has given a lot of money to farmers in the east of Germany to buy much needed machinery, equipment and materials for new buildings.

Farmers in the east grow potatoes, barley and wheat. They also rear cattle, pigs and poultry.

The environment

Modern farming methods and the chemicals that are sprayed on crops are damaging the environment.

Hedges are being cut down to make way for fields that can be used as farmland. The hedges are home to insects, birds and lots of small animals. ▼

▲ *This farmer is spraying pesticides on his crops to kill insects.*

Most farmers in Europe use chemicals to make their crops grow more quickly.

The chemicals contain a fertilizer called nitrate. Sometimes the nitrate is washed into streams and rivers. It pollutes the water and sometimes poisons the fish.

Scientists are worried that nitrate which gets into our drinking water can make us very ill.

Lake Prespa

Lake Prespa is in the north of Greece. The beautiful lake is being destroyed by modern farming methods.

Lake Prespa is home to many rare birds. A lot of them are killed by fish farmers. This is because the hungry birds eat fish from the fish farms on the lake.

The lake is polluted by food and waste from the fish farms. Pollution also comes from chemicals in the water that drains into Lake Prespa from nearby fields.

▲ *This sleek, proud animal is a lynx. Sadly, lynx are becoming extinct.*

Farming is also a threat to wildlife. Farmers need more land in order to produce more food. The land that is being ploughed up and sprayed with chemicals is home to all kinds of birds, animals, insects and wild flowers.

If we do not do something soon, some of the animals we see every day, like frogs and butterflies, may never be seen again.

The wetlands

▲ *Frogs live mostly in fields and swamps. This land is being drained of water and used for farming. The frogs have nowhere left to go.*

The wetlands is the name given to areas of wet, muddy swamps and marshland.

The plants, animals and reptiles that live in the wetlands have lost their homes to farmland. Many are also killed by pollution from farms.

The Camargue

The Camargue is an area of marshland in France. It makes an ideal home for mosquitos who love the wet swamps, black bulls and huge flocks of bright pink flamingos. It is famous for the wild, white horses (above) who roam free across the area.

The Camargue has the right climate and type of soil for growing rice. Land in the area has been drained and planted with rice.

The rice is sprayed with chemicals. These are polluting the water. Now the whole of the Camargue, and all the plants, animals and birds which live there are in danger.

Farming tomorrow

▲ *Can you see the roots of this tree? The soil has been eroded (worn away) by the weather. Now, this tree that has stood for over 100 years, will die.*

Farming is and always will be a major part of life in Europe. In some European countries, the new machinery and chemicals used by farmers have been too successful. Land is ploughed up and used to grow more crops. Yet these crops are often destroyed.

▲ *Sometimes it is easy to think of today, of ourselves and our needs. We must think of the environment and the future. We must make sure that our farming methods do not destroy the world we live in.*

Many of the countries in Europe are trying to take back some of the land that has been used for farming. Instead the land is planted with trees or used as parks for tourists.

Farmers in Europe are being asked to produce more food, yet this is damaging the environment. There is not enough food in many countries in the world. However, some countries in Europe produce too much.

As Europe faces the future, there are many decisions to be made.

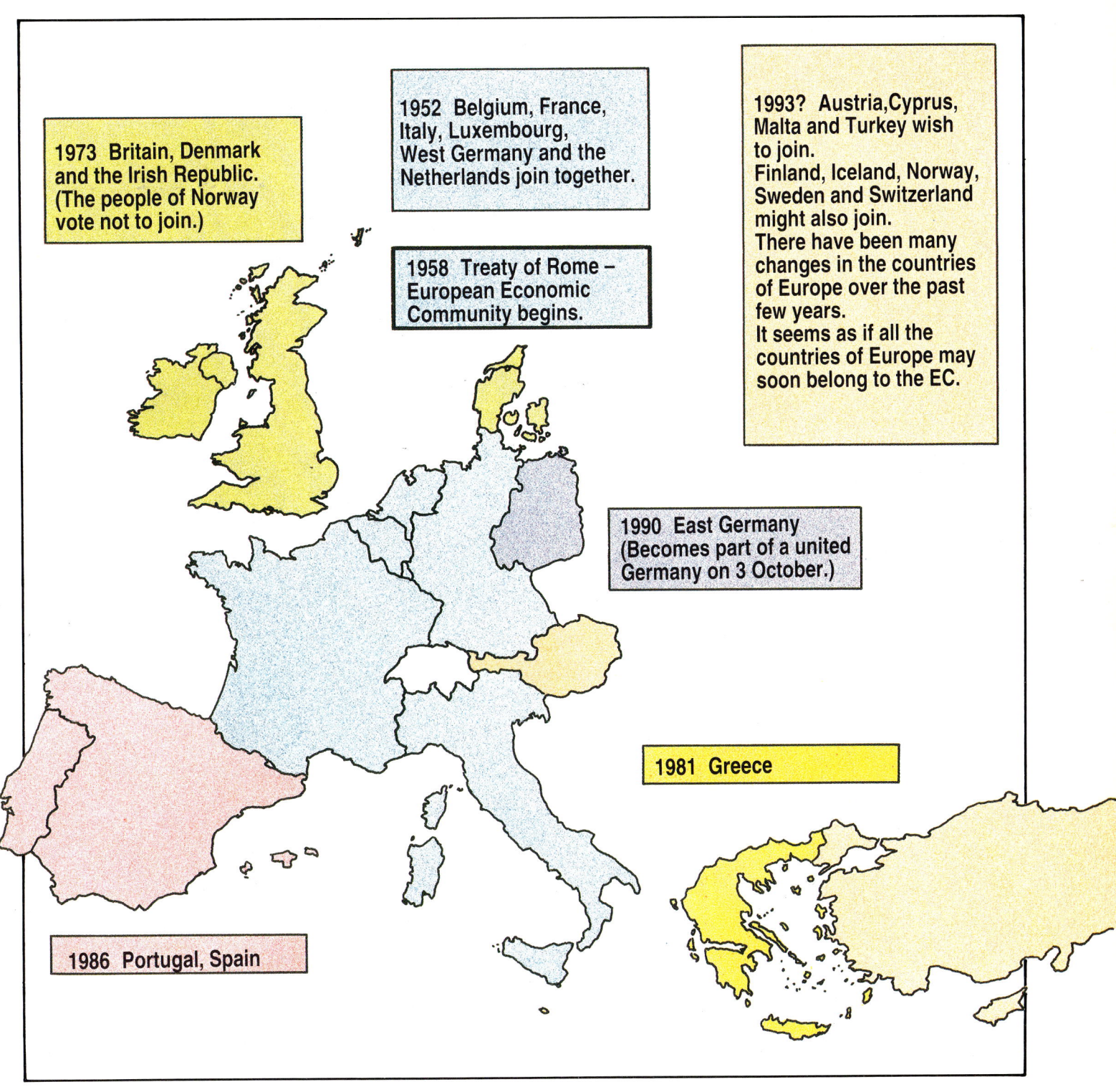

1973 Britain, Denmark and the Irish Republic. (The people of Norway vote not to join.)

1952 Belgium, France, Italy, Luxembourg, West Germany and the Netherlands join together.

1958 Treaty of Rome – European Economic Community begins.

1993? Austria, Cyprus, Malta and Turkey wish to join.
Finland, Iceland, Norway, Sweden and Switzerland might also join.
There have been many changes in the countries of Europe over the past few years.
It seems as if all the countries of Europe may soon belong to the EC.

1990 East Germany (Becomes part of a united Germany on 3 October.)

1981 Greece

1986 Portugal, Spain

▲ *This map shows the countries of the EC and the year they joined.*

Glossary

Churn   Shaking milk or cream to make butter.

Cotton bolls   The round flowers of the cotton plant which hold the seeds.

Environment   The world around us. For example, animals, plants, rivers, mountains and the air we breathe.

Extinct   No longer living on the Earth.

Fertile   Able to make things grow.

Fertilizers   Chemicals which are sprayed on crops to help them grow.

Glaciers   Huge blocks of ice.

Graze   To feed on grass in a field.

High-tech (Short for high-technology.)   Very up-to-date machines and methods.

Irrigation ditches   Narrow trenches in the ground that collect and store rain-water. The water then passes along pipes to fields.

Nature reserves   Large areas, looked after by the government, where plants and animals can live and grow safely.

Percentage   The number out of every 100. For example, 10 per cent means 10 out of every 100.

Pesticides   Chemicals which are used to kill insects that live on crops.

Polder   A stretch of land which was once covered by sea-water. The land is now used for farming.

Policies   Ideas and ways of doing things that have been agreed on by a group of people.

Pollutes   Spoils the environment and harms animals.

Produce   The crops or animals raised by farmers.

Rare   An animal or plant is rare when there are not many of its type left in the world.

Refining   Getting rid of all dirt and unwanted substances.

Reptiles   Cold-blooded animals with scaly skins, such as snakes and lizards.

Traditional   Has been done for a long time.

Udders   The parts of a cow which hold milk.

Vineyard   Large fields where grapes are grown and later turned into wine.

More information

Books to read

Dairy Farming by Joy Palmer (Hodder & Stoughton, 1989)
Farming by Sue Hadden (Wayland, 1991)
Food and Farming by John Becklake and Sue Becklake
  (Franklin Watts, 1991)
Modern Farming by Jeff Battersby (Franklin Watts, 1990)
Our Country series (Wayland, 1991-2)

Useful addresses

If you would like to find more information on farming in Europe, you
can write to these organizations.

Commission of the European
  Communities
8 Storey's Gate
London SW1P 3AT

Council of Europe
Boite Postale 431 R6
67006 Strasbourg Cedex
France

Friends of the Earth
26-28 Underwood Street
London N1 7JQ

Index

Albania   28
Alps, the   10-13
Atlantic Ocean   4
Austria   10

Barley   14, 33, 34, 36, 37
Belgium   30
Bosnia   28
Bulbs   19
Bulgaria   28
Butter   10, 12, 31, 34

Camargue, the   42
Cattle farming   9, 10, 12, 19,
   24, 36
Cheese   10, 12, 16, 34
Chemicals   31, 38, 40, 42, 43
Chickens   27, 33
Collective farms   25, 28
Combine harvesters   27
Common Agricultural Policy
   (CAP)   30, 31, 32, 37
Cotton   6, 22
Cream   34
Croatia   28
Czechoslovakia   28

Dairy farming   10, 11, 12, 16,
   24, 31, 32
Denmark   34-5
Dykes   16, 17, 18

Eastern Europe   25-9
Environment, the   6, 14, 31,
   38-40, 41-2, 44
European Community (EC),
   the   30-37, 45

Fertilizers   23, 24, 30, 34, 35,
   39
Figs   21

Finland   12
Fishing   4
Fjords   14, 15
Flowers   14, 24
France   10, 33, 42

Germany   36-7
Glaciers   4, 10
Goats   9
Greece   6, 9, 21, 22, 40

Hay   10, 13
Hungary   28, 29

Iceland   4
Ireland   32
Irrigation   20, 22, 23
Italy   9, 10, 21, 22, 23, 24

Machinery   24, 28, 29, 30,
   33, 35, 43
Maize   21, 33, 36
Marshland   41
Mediterranean Sea, the   9
Mediterranean, the   9, 20-24
Methods of farming
   modern   22, 24, 33, 35, 37,
      40
   traditional   22, 23, 33
Milk   10, 11, 12, 26, 31, 32,
   34, 35

Nature reserves   18
Netherlands, the   16-19
North Sea, the   18
Norway   14, 15

Olives   4, 9, 21, 33

Pesticides   34, 39

Pig farming   12, 26, 27, 34,
   35, 36, 37
Poland   25, 28
Polders   14
Pollution   39, 40, 41, 42
Portugal   4, 21
Potatoes   14, 19, 21, 36, 37

Raw materials   6
Reptiles   41
Rice   21, 42
Romania   28
Russia   25-7

Serbia   28
Sheep farming   9, 26, 31, 33
Spain   4, 9, 21, 22
State farms   25, 28
Sugar
   beet   6, 14, 19, 26, 36
   cane   6
   refining   6
Sweden   14
Switzerland   10, 12

Ukraine, the   25, 27

Vegetables   14, 19, 21, 23,
   27, 33
Vines   21, 33

Weather (climate)   7, 9, 14,
   21, 23
Wetlands, the   41-42
Wheat   14, 19, 33, 34, 36, 37
Wildlife   31, 40
Wine   21, 22, 36

Yugoslavia   10

Zuider Zee, the   18